THE MAJESTIC DINOSAUR
A History of the Middlesbrough Transporter Bridge

Martin Phillips

Lyndhurst Publications ◆
◆
◆

About the Author

Martin Phillips who is married with two young daughters is Head of History at Four Dwellings School in Birmingham. He was born in Middlesbrough in 1968 and went to Linthorpe and Boynton (now Hallgarth) schools before attending Acklam 6th Form College. He then gained an honours degree in History and Politics at Leicester University. As well as local history, he keeps up his interest in sport especially playing five-a-side football and, of course, watching the Boro at the Riverside whenever possible.

LYNDHURST PUBLICATIONS LIMITED
1 Belasis Court,
Belasis Hall Technology Park,
Billingham, Cleveland. TS23 4AZ

First Published July 1992
2nd Edition November 2006

Copyright © 1992 and 2006
Lyndhurst Publications

ISBN 1-874766-02-9

Typeset in Palatino and printed
By Mags Laser,
Middlesbrough, Cleveland.

CONTENTS

ILLUSTRATIONS

FOREWORD

BY NEIL KENLEY

This second edition of the history of the Middlesbrough Transporter Bridge brings the story right into the twenty-first century. Nearly a hundred years ago the Bridge was at the leading edge of engineering technology, and this edition finishes with a look forward to how it is has been used as a test-bed for the new clean hydrogen fuel-cell technology.

That the Middlesbrough Bridge is one of only a handful of working transporter bridges in Europe is a living testament to the quality of the conception, design and construction by the people who planned and built it. All these qualities are just as alive now in the Tees Valley. From the top of the Bridge one can view scientific innovation being turned into industrial reality as massive new investments take shape, especially in the field of environmentally friendly fuels.

Its elegant commanding presence has long been an icon of Middlesbrough, where it is still reflected on the town's new logo. But the 1994 addition of blue floodlighting means that it can be easily picked out from many a vantage point throughout the Tees Valley.

This is a fascinating story of the interplay of politics and economics with science and technology. But, of course, none of these abstract things work in an abstract way. History is all about people, and there are plenty of people in this book including Terry Scott who had a lucky landing in the safety mesh, and the cast of Auf Wiedersehen Pet who put the Bridge on the international scene when in 2002 they took it (well, the TV company had to publish a notice that it hadn't really been taken!) piece-by-piece to the United States. After this series over 1,000 people visited the Bridge on Heritage Day. It also appeared in the film Billy Elliot, the TV series the Fast Show, and in a programme in the second BBC TV series of Coast in Autumn 2006.

On behalf of Tees Valley Regeneration I welcome this second edition, and look forward to our successors celebrating the second century of the Middlesbrough Transporter Bridge.

Neil Kenley
Strategic Investment & Marketing Director
Tees Valley Regeneration
Cavendish House
Stockton-on-Tees, Tees Valley, TS17 6QY

October 2006

2006: Transporter Bridge with DI Oils' biodiesel refineries, Forty Foot Road, Middlesbrough, in the foreground.

Ready for action - July 1992

THE MAJESTIC DINOSAUR

A HISTORY OF THE MIDDLESBROUGH TRANSPORTER BRIDGE

1   INTRODUCTION

To the people of Middlesbrough and Cleveland the Transporter Bridge is their paramount symbol of home. Indeed it forms part of the modern logos of both the Borough and the County. From the day of its opening in 1911 it became the beacon of Middlesbrough's leading role in the world of industry. For any Teessider it represents all the pride of our industrial heritage, and today again provides the vital link across the Tees for thousands of Cleveland's workers.

The sight of the Bridge is always welcome to the eye of a returning exile, and there are many such folk today. Over eighty years after its opening the Bridge awaits its return to full-time operation, an event that will revitalise the crossing and illustrate new hope for the people of Middlesbrough.

In his book about the Bridge, Alan Dowson referred to a critic of the Bridge who described it as "looking like the fossilised skeletons of a couple of prehistoric pterodactyls' heads bent aggressively over the Tees." The prehistoric allusion need not be seen as critical. In fact the comparison to a dinosaur may well be an apt one for it conjures up an image of enormity and strength: omnipresence and longevity: an all-surveying creature of expansive power.

Unwittingly, I believe this critic to have provided an inspirational rather than a critical comment upon this monument to Cleveland's engineering spirit and achievement.

2 ORIGINS OF A CONCEPT

The first proponent of the transporter bridge concept was the Hartlepool engineer, Charles Smith. He submitted his design for the Middlesbrough Bridge to the Borough Council as early as 1873 when it was desired that a regular and efficient crossing be put in operation between Middlesbrough and Port Clarence. His quote for the scheme was £31,162 plus an added £1,260 to cover maintenance and operation based on a two-shift system.

Despite the backing of an eminent bridge authority, Sir Benjamin Baker, his idea was rejected. The accepted tender was that of Messrs Hodgson and Ridley whose proposal to build landings for a horse and cart ferry struck a more familiar and less expensive tone.

In fact it was a third of the estimated price of Smith's proposal and this fundamental factor - the cost to the local purse - would continue to be the Bridge idea's greatest obstacle. The ferry system would remain and be operated in various guises before the Bridge opened in 1911.

Smith, though his tender was rejected and his designs never reached the construction stage, has been accredited as being the inventor of the transporter bridge. However it was Monsieur Ferdinand Arnodin who first transformed idea into reality. He pioneered the idea by constructing bridges in Bilbao in Spain and Nantes and Rochefort in France. His expertise was to become a significant influence behind the conception of the Middlesbrough Transporter Bridge.

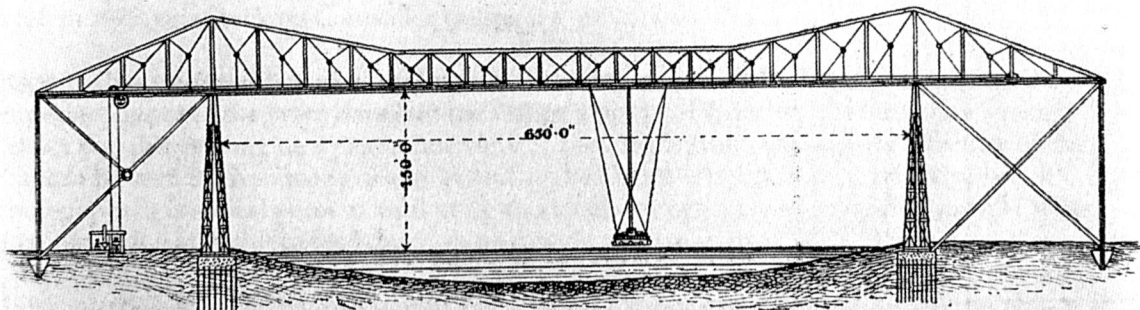

Proposed Ferry Bridge across the River Tees at Middlesbrough in 1873
Designed by Mr. Charles Smith, Engineer, Hartlepool

1

ALDERMAN J. McLAUCHLAN
(Pioneer of the Transporter Bridge Scheme)

Alderman J McLauchlan

3   THE BIRTH OF AN IDEA

After Smith, the earliest advocate of a Middlesbrough Transporter Bridge was Alderman Joseph McLauchlan. He had visited the transporter bridges already in existence on the Continent and believed the design to be appropriate for Middlesbrough's needs. In a letter to the Ferry Committee in June 1901 he drew attention to the widely acknowledged need to update the crossing facility over the Tees between Middlesbrough and Port Clarence. He reminded the Committee that during recent years various bridge and tunnel schemes had been suggested but none had been offered any credibility or debate because of the potential cost to the Corporation, and so the ferry service remained.

McLauchlan, who later became the town's Mayor, referred to the transporter bridges designed by Arnodin in Rouen, Bizerta, Bilbao, Rochefort and Nantes. Describing the bridges' main features he advanced the potential advantages such a design could offer to this particular crossing: the site was ready made as the road approaches to the present crossing were well-established; the cost of operation would be cheaper than the present ferry system as it only required the services of one man at a time and much of the cost would be covered by an increase in revenue which would result from a greater passenger capacity. He added that the two boats presently operating, the 'Hugh Bell' and the 'Erimus', would shortly need replacing after twelve years service, at a cost of £12,000. McLauchlan estimated that the cost of such a bridge would be approximately £40,000. His assertion was supported by the Westminster-based engineering consultant Charles H. Gadsby, and McLauchlan called for a detailed report. Gadsby had acted as consultant to the scheme to build a similar bridge across the Tyne to link North and South Shields. The Shields Bridge Bill had been defeated in Parliament because of its potential vulnerability in time of war, and the effect its destruction would then have on local industrial production. The report sought by McLauchlan was not forthcoming but McLauchlan had succeeded in sowing the first seeds of interest. Two representatives of the Mersey Transporter Bridge project linking Widnes and Runcorn wrote to the Council offering their advisory services and to compile a report.

The Ferry Committee were at this stage ambivalent towards the proposal and did not develop McLauchlan's initiative further but the issue certainly did not evaporate. The subject of the ferry service remained at the forefront of Council affairs as confidence in it began to deteriorate. On 17th December 1901, the Committee was presented with two letters from employers north of the river. Greville T. Jones, manager of the Clarence Iron Works and Edward Dawson, managing director of the Anderston Foundry Co. Ltd., raised concern about the ferry's reliability. Jones commented that when the 'Hugh Bell' was not running its replacement was often too small and dangerously unsuitable for carrying 170 passengers. In one particular instance, stormy conditions delayed the ferry and as a result many of his employees were late; the ironworks could not commence work until eight o'clock, and there was a 35% drop in the day's production level. The loss of earnings meant the men were suffering financially due to the 'general mismanagement' of the ferry which was often inoperable in bad weather.

Certainly the service was by no means the epitomy of safety. The Council proceedings in the years up to 1911 are littered with references to accidents incurred on ferry crossings.

One such accident involved a local butcher's delivery man. As he approached the ferry in his cart the boatman pulled up the ferry door and the horse, unable to stop, carried the cart and the delivery man into the river. His employer, a Mr. P. Milburn, made a claim for the man's lost wages, the cart, and his biggest casualty - twenty stones of beef worth £8.

Another claim was made by a mother whose son lost a toe-nail in an accident involving the ferry's swing door. Due to the consequently enforced absence, he lost seven week's wages which was an enormous loss to a household reliant on his income. The council admitted liability in neither case and both claims for compensation were rejected.

Such financial suffering was not confined to passengers. A petition dated April 4th 1904 was sent to the Ferry Committee from the ferry operators. All but one had worked on the ferry for over ten years and two of the men had been there for twenty-eight. Despite a large increase in the river traffic they had not received a pay rise for over four years and appealed for a pay review. The Committee's decision? 'Ordered that the application be not entertained.'

4   THE IDEA GAINS WEIGHT

These issues all highlighted the unquestionable need to update the crossing facility which in its present state was poorly run, unreliable and often dangerous. Recognition of this was made apparent in June 1904 when the Ferry Committee authorised the formation of a sub-committee, headed by McLauchlan, to consider the suitability of a transporter bridge as a means to link Middlesbrough with Port Clarence. Later, in September of that year, the Committee gave its approval for a deputation to visit the transporter bridges in operation at Rouen and Nantes. This was before the days of twin towns and paid visits however, and any expenses were to be met by the Councillors themselves.

The sub-committee was McLauchlan's first breakthrough but the idea did not gain any weight until a special Council meeting on March 6th 1906. The Middlesbrough and West Hartlepool Light Railway Company had applied, in November 1905, for permission to build a transporter bridge over the Tees in order to link existing tramways with their railway to West Hartlepool. It was now a major issue. Alderman Amos Hinton proposed, and Alderman Archibald seconded, 'that the Corporation continue to offer the most strenuous opposition to the scheme of the Promoters as far as the Transporter Bridge is concerned.'

The meeting marks the turning point in this stage of the Bridge's history. The Council was now committed, in effect, to adopting the idea of the Bridge. This resolution was supported by a petition read to the Committee exactly a week later to the effect that the town's ownership of the crossing should not be jeopardised and therefore if a bridge had to be constructed it must be run by the Corporation rather than by an 'outside speculative company'.

The Light Railway Company had forced the hand of the Council but 'speculative' it may well have been. One cannot fail to question whether the application, which was given permission supended for one year, was indeed made with serious intent or merely for the purpose it did in fact achieve. It certainly created an advantageous scenario for proponents of such a construction and left the Corporation with no room for maneouvre if it was to retain authority over the public crossing to which it was determinedly committed.

The affair illustrated the increasing range of responsibilities accepted by local government during this period. It also demonstrated the pace of technological advance. In 1903 Orville Wright made the first power driven flight in an aeroplane; in just over a decade it was to become a machine of war. These were rapidly moving times and the Corporation had to keep up.

It must be appreciated that this apparent conservatism demonstrated what a modern development the Transporter Bridge was in terms of engineering technology; the concept was too advanced for some Councillors to accept. Reports and cartoons appeared in the local press referring to the Bridge as the 'Flying Ferry'. Councillor Allison described the Bridge scheme as a 'showman enterprise' with a sensational element more befitting a private company. He commented sarcastically that he "would sooner support a scheme for an airship." Similarly Councillor Sadler believed it would be throwing money into the wind and expressed concern at the ratepayers' burden whilst Councillor Mascal's fundamental opposition was that he was not convinced it would be the best form of crossing.

Hinton and McLauchlan had allies in arguing that the speed of a transporter bridge 'would remedy the difficulty workers had in arriving to work on time using the present system'. Prompting laughter in the Chamber, Councillor Loughran said that: "It was all very well for Alderman Archibald to speak of crossing in Noah's Ark (a popular term for the ferry) but those who used Noah's Ark didn't have to keep time with the buzzers."

The safety aspect was a further factor of support. Under the ferry system, if anyone was injured on the north side of the river at night someone had to shout across the river to alert the boatman, who may even be asleep at home, in order to ferry the casualty over to get to the infirmary. It was argued that a nightwatchman would alleviate this not infrequent problem.

Councillor Mattison stated that there was evidence that it could be afforded; it could increase revenue and in the long-term would be much cheaper to run than the boats. The success of those who championed the proposal was largely due to their vigour in acquiring financial costings and forecasts in order to overcome what had always been the backbone of opposition to the Bridge - cost.

4

5 DIVISION IN THE TOWN HALL

The Borough Council remained deeply divided over the issue. It is ironic that Alderman Archibald should support Alderman Hinton's early proposal of 'strenous oppositon' as the two men later became leading personalities on opposite sides of the Bridge Debate that ensued. Hinton became actively pro-Bridge and produced costings that showed that there was no threat of such a scheme causing significant rate increases. Indeed he advanced that the scheme could have the opposite effect. He based his financial argument on a bridge producing at least the annual revenue of the present ferry, £7,000 p.a.   The interest and redemption covering the construction could be paid at 5% over a forty year period at a cost of £3,000 p.a. Other charges such as maintenance, operating costs and taxes would come to £1,600 p.a. leaving a contingency margin of £2,400. Given that such a bridge would inevitably raise a greatly increased level of revenue the scheme was beginning to look affordable even if the basis for Hinton's estimates were unclear. He also promoted the idea of a public vote on the scheme which he was confident he would win.

Alderman Archibald was a determined opponent of the idea which he dismissed as a passing notion. He believed that the boat service would be satisfactory for another 20 to 25 years and shared with Councillor Dorman a concern for the futures of the ferrymen and the owners. He stated that no construction plans had been drawn up nor could anyone categorically state how secure the foundations would be. In response the sub-committee inquired into the prices of necessary borings, which were taken in the October of 1906.

6 DEVELOPMENT - TOWARDS CONSTRUCTION

Progress was now being made, steadily if not rapidly, towards the construction of the Bridge. Expert advice was sought and Mr. A.C. Pain, a London engineer, produced a report in July 1906 which confirmed the view that a transporter type bridge would be the most suitable means of crossing the Tees at this point. The traffic along that part of the river was much too dense for either a swing, draw or bascule bridge to work without causing undue inconvenience. The cost of any such project or that of a tunnel was also simply beyond consideration. The ferry service had proved itself to be outdated and therefore only a transporter bridge would be appropriate.

On the 5th October 1906 it was agreed to recommend:
1. That a Transporter Bridge be erected over the River Tees and that the Corporation land at each side be utilized for the purpose.
2. That the Town Clerk be instructed to take all the necessary steps for the promotion of a Bill in Parliament for the purpose of obtaining the requisite powers therefore.
3. That the Town Clerk engage Parliamentary Agents to act in conjunction with him.
4. Adjournment - Ordered that this meeting be now adjourned to consider the question of the appointment of an expert engineer etc.

It was now official policy and the Council were to put the Bill to Parliament. The Bill and legal notices were drawn up by Mr. Preston Kitchen the Town Clerk of Middlesbrough. At a special meeting later that month it was agreed that the necessary notices would be prepared and the question be discussed with Monsieur Arnodin. Also confirmed was the site of the Bridge; to pursue other options or prepare another site would incur needless expense. A special meeting in December 1906 confirmed that the Bill, then in its draft stage, would be promoted, that the ferry would be discontinued once the Bridge was built, and that the Bridge would be funded by local money and rates.

On 8th January 1907 a public meeting was held in the Middlesbrough Town Hall explaining the plans for the Bridge. On 24th January the public were invited to vote on the project. The result was 2,255 in favour with 1,620 voting against the motion. Amos Hinton's prediction was vindicated by a majority of 635 votes.

Certainly there was strong opposition to the scheme but a resolution proposed by Hinton demonstrated a pragmatic policy development. The resolution was put to the meeting and carried: 'Ordered that in the event of the refusal by the Commissioners of the Board of Trade to grant an order to construct a Transporter Bridge to the Promoters this Corporation will undertake to make application as early as possible for power to improve the means of conveyance between Middlesbrough and Port Clarence by the construction of a Transporter Bridge or other equally rapid and efficient means.'

In short, if the Light Railway Company failed to gain Parliamentary approval for such a construction then the Council would accept responsibility to do so themselves.

Interest in the arrangements continued to spread and in January 1907 the Ferry Committee received a letter from V.J. Kerihuel, the resident engineer of the Newport Transporter Bridge in Monmouthshire (now Gwent), offering to take up the same position in Middlesbrough. Several such letters arrived from engineers around the country who had recognised the project's potential.

The Bill was presented to Parliament and received the Royal Assent on 4th July 1907. Mr. G.C. Imbault, Chief Engineer of Cleveland Bridge and Advisory Engineer to the Corporation could now proceed with his designing of the Bridge.

In January 1909 firms were publicly invited to tender for the contract to build the Bridge, with Cleveland Bridge & Engineering Co. Ltd. having been appointed to supervise its design. Six tenders were received and on 20th April it was ordered that the one submitted by Sir William Arrol & Co. of 85 Preston Street, Glasgow be accepted. The contract was signed on 9th June 1909 for £68,026 6s. 8d. The firm began work on the immediate site on 2nd July. Arrols were given 27 calendar months to complete the construction of the Bridge.

The formal laying of two foundation stones was made on the Middlesbrough side on 3rd August 1910. The ceremony was performed by Lieutenant-Colonel T. Gibson Poole, the Chairman of the Ferry Committee, and, appropriately, Joseph McLauchlan.

7 THE MEN WHO BUILT THE BRIDGE

Monsieur Ferdinand Arnodin of Chateau Neuf sur Loire was the world's pioneer of the transporter bridge system. He had designed the first such bridge to be constructed in Bilbao in 1893. He received a Council delegation in France and visited Middlesbrough during the initial stages. His advice on the Middlesbrough proposal was significant and his approval was crucial to progress. A colleague of his throughout the earlier successes was another Frenchman, Mr. G.C. Imbault, who was now the Chief Engineer of the Cleveland Bridge & Engineering Co. Ltd. Imbault, having supervised the building of the Newport Transporter Bridge, became the Advisory Engineer to the Corporation and was to be the designer of the Bridge.

Under William Pease, Cleveland Bridge had achieved an eminent position in world engineering. Established in 1877, the company forged its reputation building bridges, industrial premises, power stations, pipelines and dock gates. The success of the Transporter Bridge was to enhance this reputation, and Cleveland Bridge with Dorman Long went on to build the Auckland Harbour Bridge. Greater still were projects completed in conjunction with Dorman Long and William Arrol - the Severn Bridge and the Forth Road Bridge. More recently Cleveland Bridge constructed the A19 Flyover across the Tees and have been leading contractors in the Canary Wharf and Thames Barrier projects and the new Hong Kong Bridge.

William Arrol & Co. Ltd. of Glasgow had made much of their reputation in the 1880s and 1890s. They reconstructed the Tay Bridge, built the Forth Railway Bridge and were the main contractors for London's Tower Bridge which was completed in 1894. Most of the labour involved in the Middlesbrough Transporter Bridge was local and operations were carried out under the supervision of a director of Arrol's, A.S. Biggart and on behalf of Middlesbrough Borough Corporation, S.E. Burgess, the Borough Engineer. Robert Anderson assumed the position of Resident Engineer. H.M. Taylor, the Borough Electrical Engineer, supervised all electrical work on the Bridge.

Sir William Arrol's workforce - 4th February 1911

7

Mr. G. C. IMBAULT
of the Cleveland Bridge Co., Ltd., Darlington, Designer of the Bridge

Mr. A. S. BIGGART
of Sir Wm. Arrol and Co., Ltd.

Mr. R. C. MACDONALD
Borough Engineer, Upon Wm. Arrol and Co., Ltd.

SIR WM. ARROL
Contractor

Mr. ROBT. ANDERSON
Resident Engineer for the Middlesbrough Corporation

Mr. S. E. BURGESS
Borough Engineer, acting superintendent the Erection of the Pier and Span of the Bridge

Mr. PRESTON KITCHEN
Engineer

8

8 THE DINOSAUR GROWS

In 1961 at the Fiftieth Anniversary celebrations, Preston Kitchen recalled how he had been the first to walk along the full length of the walkway. The span had not been fully joined and, encouraged by Mr. Ennis the Bridge Engineer, he jumped over the gap of about a foot.   Mr. Ben Chivers also had proud recollections at the Anniversary celebrations. He had been the first to test the strength of the car platform by driving a steam-roller onto it.

What is pleasing to report, bearing in mind the working conditions and the primitive health and safety regulations we would associate with that period, is the surprising fact that nobody was killed while working on the construction of the Bridge.

The following set of photographs shows the working conditions during the foundation stages of the Bridge's construction.

Completed wing wall on the Middlesbrough side - 3rd June 1910

9

Retaining wall apron and bottom boom base girder on the Middlesbrough side - 5th July 1910

The two towers rising on the Port Clarence side - 1st August 1910

The cantilever progresses on the Middlesbrough side - 11th March 1911

The cantilever progresses on the Port Clarence side - 11th February 1911

CENTRE OF PIN LEVEL 270'-0" A.O.D.

STAIRCASE

POWER HOUSE
BASE LEVEL AT
CENTRE 116·85

HIGH. WATER LEVEL 108·8

MIDDLESBROUGH SIDE

70'-0" BELOW H.W.

90'-0" BEL

140'-0"

571' 0" BETWEEN TOWERS

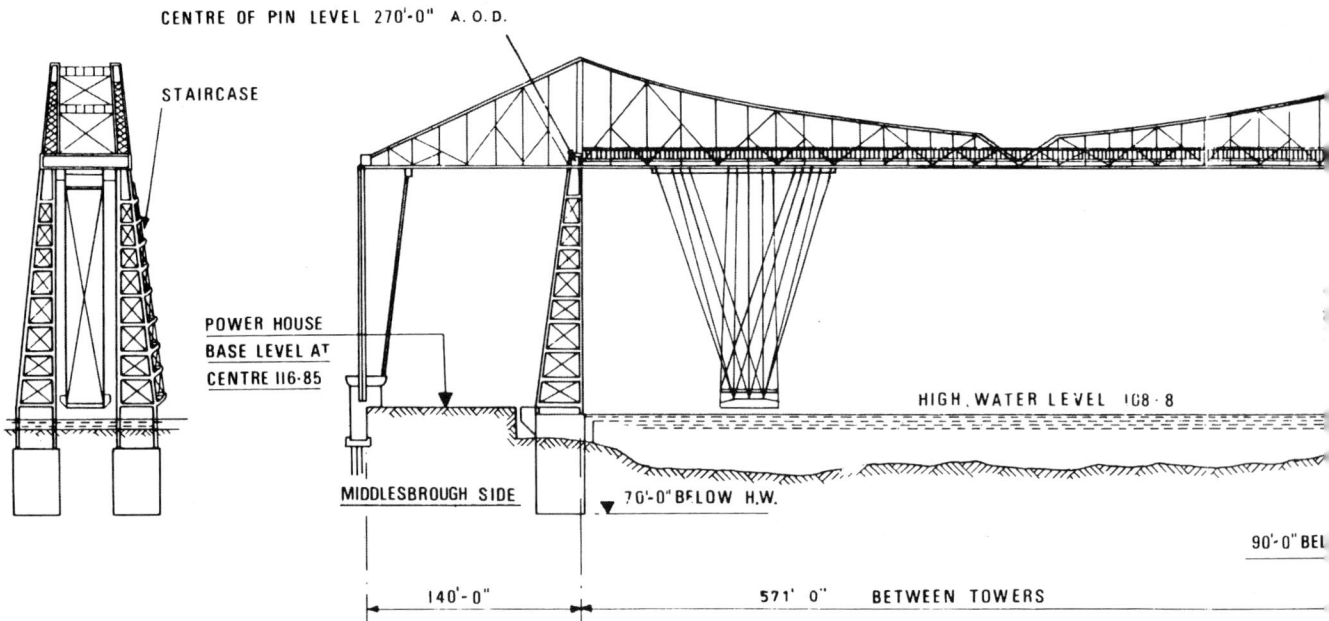

CROSS SECTION

GENERAL ELEVATION

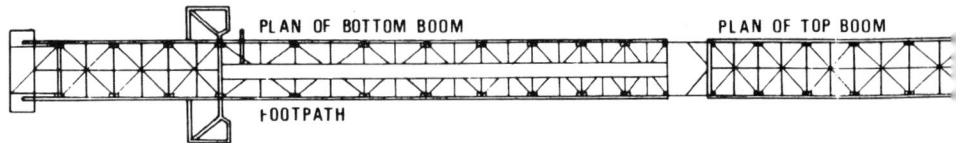

PLAN OF BOTTOM BOOM

PLAN OF TOP BOOM

FOOTPATH

PLAN

MIDDLESBROUGH
1911 – 1968

TEESSIDE
1968 – 1974

13

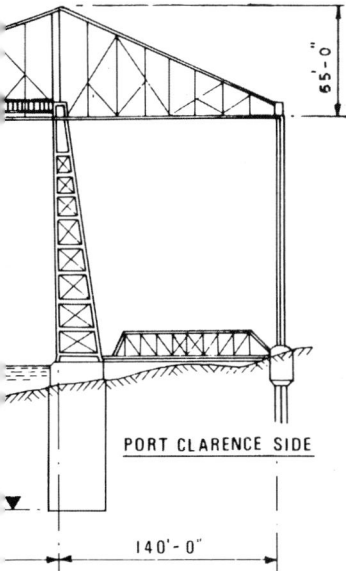

PORT CLARENCE SIDE

140'-0"

65'-0"

35'-0"

Total length of the bridge and its approaches. . . . . 850 feet
Span between centre of the towers. . . . . 571 feet
Clearance from low water mark to the
underside of the girders. . . . . 177 feet
Extreme height of the bridge from high water mark
to the top of the centre of the towers (i.e. to the top
of the main posts of the cantilever girders) . . . . 225 feet
Length of each base girder on which the towers
are built and which are supported on concrete foundations. . . . 98 feet
Depth of the base girders on which the towers
are supported. . . . . 16 feet
Weight of the base girders . . . . 163 tons
Foundations have been sunk to a depth below high water of :-
Port Clarence side . . . . 90 feet
Middlesbrough side . . . . 70 feet
Steel caissons are sunk to these depths and the towers
for the support of the superstructure of the bridge
are built thereon.
Caisson foundations are filled solidly with concrete.
Total quantity of concrete used in the foundations
including retaining wall supports . . . . 10,000 cub. yds.
Total amount of steelwork in the bridge . . . . 2,600 tons.
Steelwork in caisson foundations   600 tons

TESTED WITH 80 TONS OF PIG IRON.
GUARANTEED TO CARRY 860 PASSENGERS.
600 WITH TRAMCAR AND PASSENGERS.

NOTE:- Datum 100 feet below O.D.
        8657.
        EM/31.
        Opened   October 17th.,1911.

CLEVELAND COUNTY
1974 —

County of Cleveland				
Department of the County Surveyor & Engineer				
TRANSPORTER BRIDGE AT MIDDLESBROUGH.				
PLAN, ELEVATION, AND CROSS SECTION.				
Scales	Hor.	Drawn		G.M.W. DRABBLE
	Vert.	Traced	R. SMITH.	B.Sc., Dip.T.P., C.Eng., F.I.C.E. F.I.H.T. F.R.S.A.
Date	28·11·75	Checked		COUNTY SURVEYOR
DrgNo	HB/M 3112/1	C 297		& ENGINEER

14

I April 1911

The meeting of the pterodactyls. General views as the cantilevers march towards the centre.
The 9th May 1911 - 'a good days work!'

II 9th May 1911

III 9th May 1911

IV 16th June 1911 - note the height of the three master

Trial test - early autumn 1911
The look of apprehension is evident on many of the faces on both the trial and opening runs

The Official Opening run - 17th October 1911
The Prince is seen standing on the left-hand side of the podium

9 THE OPENING OF THE MIDDLESBROUGH TRANSPORTER BRIDGE

The Opening Ceremony was one of the biggest events Middlesbrough had witnessed. Shops, factories and schools closed for the day as the local public made their way to the riverside to watch the formal occasion. To a guard of honour of the local Crimean War Veterans, H.R.H. Prince Arthur of Connaught arrived at 10.30 on the morning of 17th October 1911. " The Royal Visit to Middlesbrough was an unqualified success. The arrangements were admirable. All the events dovetailed with remarkable ease. There was no dragging. Everything happened at the appointed moment. The order preserved in the streets was perfect."

This extract from a huge four page report in the 'Tees-side Herald' of Saturday October 21st 1911 describes, as fully as is required, the organisational qualities of the event. However the occasion was more than an efficiently stage-managed event.

The press reports suggest that the Prince offered much more than a formal recognition of a fine engineering achievement. In fact his speeches, in which he praised everyone who had been associated with the Bridge's construction, " revealed not only an intellectual interest in the work before him, but a breadth and independence of view."

Indeed the Prince displayed in the speeches his links with and affection for the town. His late uncle, King Edward VII, had opened Middlesbrough Town Hall and Albert Park was formally opened by his father the Duke of Connaught. He visited Albert Park after the ceremony to plant a tree beside the central path close to the Boer War memorial.

Photo by W. and D. Downey, London]

H.R.H. PRINCE ARTHUR OF CONNAUGHT, K.G.

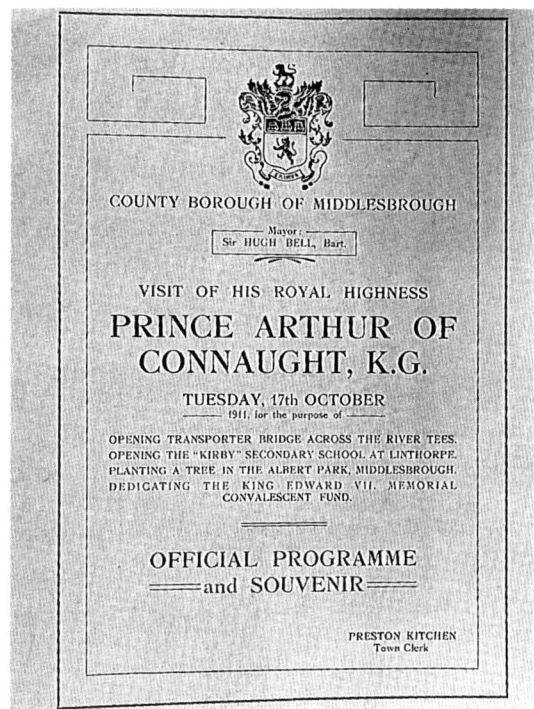

COUNTY BOROUGH OF MIDDLESBROUGH

Mayor:
Sir HUGH BELL, Bart.

VISIT OF HIS ROYAL HIGHNESS

PRINCE ARTHUR OF CONNAUGHT, K.G.

TUESDAY, 17th OCTOBER
1911, for the purpose of

OPENING TRANSPORTER BRIDGE ACROSS THE RIVER TEES.
OPENING THE "KIRBY" SECONDARY SCHOOL AT LINTHORPE.
PLANTING A TREE IN THE ALBERT PARK, MIDDLESBROUGH.
DEDICATING THE KING EDWARD VII. MEMORIAL
CONVALESCENT FUND.

OFFICIAL PROGRAMME
and SOUVENIR

PRESTON KITCHEN
Town Clerk

The Prince then conducted the formal opening of Kirby School in Roman Road, Linthorpe. The school had cost £7,550 and hoped to enroll 263 students. In his speech he stated his conviction that education should not merely provide the basis for examinations but to prepare the young for life itself. He wanted to see the natural individual character and ability of the young developed to the full. Young people, he said, needed guidance in order to channel their natural tendencies especially as "the tendency of complicated official machinery is to squeeze out the character and to neglect the development of natural ability."

Whether he was referring to the ever-advancing machinery of heavy industry or the expanding machinations of modern bureacracy is unclear but his comments would be equally apt about both. He was especially concerned that education should emphasise the teaching of modern languages which was particularly fitting to Middlesbrough - a centre of international commerce. In 1992 these comments are as appropriate as ever and one could assume that the same speech could readily be used by any visiting Royal today.

The Prince's visit continued with a completely unrehearsed visit to the Temperance Hall which was managed by the local MP Mr. Perry Williams and his wife. There the Prince took great delight in sharing afternoon tea with the hall's 300 or so orphaned or handicapped children.

That evening he enjoyed a banquet in his honour at Middlesbrough Town Hall hosted by the Mayor, Sir Hugh Bell, and attended by hundreds of notable guests. Among them were many familiar names including Mr. and Mrs. Dorman, Mr. and Mrs. Samuelson, Lord Barnard, the Marquess of Zetland and Lady Zetland and Lady Arrol whose husband, Sir William Arrol, had sadly passed away before the Bridge's construction was completed. Foreign emissaries based in Middlesbrough were also present as were the jounalists of all the northern newspapers, the 'Daily Mail' and shipping and engineering publications. The Prince again took pleasure in an occasion of great hospitality.

The Prince seemed to greatly enjoy his comprehensive visit in which he displayed a compassionate understanding of the achievements and needs of a growing industrial town when Middlesbrough's world standing was approaching its height.

THE MAYOR (Sir Hugh Bell, Bart.)

10 REACTION / IMPACT

It has to be said that initial reaction to the Bridge was mixed. Criticism concerning its appearance was captured by the pterodactyl quotation. The local press conveyed the views of those for whom it was too extraordinary a design to comprehend. Again references were made to the 'Flying Ferry' and the cartoon featuring 'Old Father Tees' reflected the conservatism with which some observers greeted the Bridge's opening.

TIME AND THE TEES.

Old Father Tees : What changes! They will be flying across soon !!

Cartoon from Tees-side Herald - 18th October 1911
Old Father Tees : 'What changes! They will be flying across soon!!'

There was also a body of opinion that it was merely a 'modern toy' and a frivolous use of ratepayers' money. It was certainly not unreasonable for concern to be expressed at how Council expenditure was directed, and the 'Tees-side Herald' addressed this very subject. " The answer in a sentence is that Middlesbrough is and must be looking to the future, to that industrial and commercial expansion which is essential to her progress. The ceremonies...were worth all the money spent on them and more for the opportunity they afforded of bringing Middlesbrough and Tees-side prominently before the attention of the whole country."

True enough the opening put Middlesbrough in the national spotlight and gave important publicity for Cleveland Bridge. We have looked at what effect this achievement had on the company's fortunes, but it also enhanced Middlesbrough's global standing in the sphere of civil engineering: Teesside's shipbuilders continued to produce a phenomenal proportion of the world's shipping and in 1932 Dorman Long's reputation reached its zenith with the completion of the Sydney Harbour Bridge. Meanwhile transport across the river became more efficient and served to protect the prosperity of local industry and the shipping trade.

In employment and commercial terms there is no doubt that the £84,000 that the Transporter Bridge cost to build proved to be a sound investment.

The 'Evening Gazette' has, on several occasions, referred to 'Teesside's Meccano-like Transporter Bridge'. A rather warmer (and certainly more authoritative) tone is struck by Sir Nikolaus Pevsner who paid tribute to the the awesome sight of the Bridge. In his 46 volume 'The Buildings of England' he describes it as "A European Monument, in its daring and finesse - a thrill to see from anywhere."

20

11 UNDER THE DINOSAUR'S EYE

The Transporter has from its lofty view surveyed great change since it was erected.   It saw the growth of ICI's huge chemical complexes at Billingham after the First World War and at Wilton after the Second, turning nylon, Terylene and polythene into household words. The dinosaur has, sadly, witnessed the decline of the iron industry on Teesside and watched the last ship to be launched from Smith's Dock in 1986.   The old ironmasters area is empty of iron but in the distance the Transporter can see Western Europe's largest single blast-furnace at Redcar, whilst Riverside Park is now home to a multitude of new companies.   Equally encouraging is that it has seen industrial regeneration in the form of the Redpath and Davy Offshore companies providing modules for the North Sea oil and gas industries. Looking to the north the dinosaur has seen the closure of some of ICI Billingham's plants but it has also supervised the growth of Belasis Hall Technology Park as a new focal point of north east industrial investment. It has also overseen the successful redevelopment of the residential area of Port Clarence.

The bolting-down of the Newport Bridge in 1990 was a sad reflection of the decline in the volume of upstream river traffic on the Tees. That has been mainly due to the ever increasing size of modern ships such as the oil tankers and bulk ore carriers which continue to make the Tees one of Britain's busiest ports.

12 DINOSAUR IN THE PUBLIC EYE

The Bridge itself has never been out of the public eye and local media reports have regularly featured incidents concerning it since its inception.

In 1912 the manager of the Bridge, J.B. Wilson an ex-sailor, shinned to the top of the Bridge to fix new halyards to the flagpoles. His efforts were rewarded by the flagpoles blowing down shortly afterwards.

 In May 1913, a man climbed to the top of the Bridge; removed his jacket and waistcoat and plunged into the Tees. The inquest recorded that he " died from jumping from the Transporter Bridge in a spirit of bravado while under the influence of drink " - an unfortunate victim of drink diving. Several people have died jumping from the walkway and the Bridge became a location for suicide attempts. By the Bridge's Fiftieth Anniversary in 1961 three people had died from jumping. In June 1971, a man attracted attention by throwing bottles from the top. After persuasive negotiation he was talked-down by a policeman. Others were not so fortunate and the number of deaths from jumping from the Bridge was , by 1981, ten. Access to the walkway is now restricted. The stairs on the Port Clarence side have been removed and a written application must be submitted and written permission received from the Bridge Superintendent before a member of the public can walk along the top.

In 1919 a Norwegian vessel lost the top of its mast whilst passing under the Bridge and in June 1931 the North Eastern Daily Gazette carried a photograph of a windjammer, the 'Grace Harwar', passing beneath the Bridge - its masts clearing it by no more than two feet. Despite its great height this has always been a concern. During the seventies a Soviet ship was being repaired close to the Bridge. Council officials expressed concern about the enormous vessel, and enquired what its height was. Occuring at the time of the Cold War such information was withheld but the ship did manage to squeeze under the Bridge.

In 1940 a lone German plane navigated its way through the smokescreen which smothered ICI Billingham and dropped a bomb which fell through the Bridge's span and exploded on the car deck. Fortunately, it was before operation had begun in the morning so nobody was hurt, and the Bridge was open again after only three days. Similarly in 1916, Police Sergeant Booth reported seeing a bomb drop from a Zeppelin through the steel lattice work safely into the river below.

There have been several fires on the Bridge. In July 1948 firemen had to climb across to the centre of the span to fight a blaze 160 feet up which had been started by an oxy-acetylene cutting lamp falling from an upper level of staging. Two years later in August 1950, an amateur photographer was on the walkway at the top of the Bridge when he spotted a fire in one of the store cabins. He bravely ran down the stairway to alert the operators and prevent any great damage.

In April 1956 it was the car of the Bridge that was involved in a near fatality. The water-carrying vessel the Reonara collided with the car mid river and the ship's engineer Benjamin Rea was thrown into the water. Rea, a non-swimmer was plucked from the river by Fireman Robert Ferry. Meanwhile, the Reonara drifted downstream with only the deckhand, fifteen year-old Norman Duck, and his friend Leslie Walker on board. It eventually came to ground and nobody was seriously hurt though Rea and another man were treated for shock. The car was also at the centre of attention three years earlier when the Bridge broke down at high tide, leaving the car stranded for forty-five minutes halfway across the river with the car deck semi-submerged.

The history of the Transporter Bridge has certainly been far from smooth and closure-free. In 1948 the 'Hugh Bell' was brought back into service and the Corporation purchased a 36 foot long liberty-boat from the Navy during an extended closure for repairs to the running rails. In August 1958 a stoppage again put the Bridge in the headlines; the car broke down just yards from the bank and the passengers had to walk the plank to get ashore. During evening rush hour thousands of workers had no choice but to walk over the top of the Bridge. Workmen, some having worked twelve hour shifts, and secretaries wearing light summer skirts all cheerfully set about the climb. The 'Evening Gazette' reported that one gasping women said, "It was worth it for the view." In 1966 the novelty was not felt by the Middlesbrough Trades Council who complained about lost wages suffered during Bridge closures and in 1967 a petition was signed by over 200 people protesting at the inadequate contingency measures during such occasions. There seems something rather familiar about these episodes.

1977 saw a year-long period of inoperation which created considerable inconvenience, particularly when there was a simultaneous closure of the Newport Lift Bridge which prompted renewed public criticism toward the Bridge and its management. The closure of the Transporter was due to potentially dangerous fractures in the rails.

Repair notice Middlesbrough side - 20th October 1977

22

There have been incidents of people driving through the gates and into the river. In December 1952 a Bridge attendant lost control of a lorry he was driving onto the car. The lorry smashed through the gates into the river, and the driver was killed. Others to have lost their lives while working on the Bridge include a painter who fell from the structure.

A more entertaining episode featured the comedy actor Terry Scott of 'Carry On' and 'Terry and June' fame. In March 1974 Terry was returning to his hotel in Middlesbrough following his performance in 'A Bed Full of Foreigners' at the Billingham Forum. In the darkness he had expected to cross over a conventional style bridge. The carrying car was on the other side of the river and Terry drove his car over the edge, luckily landing in the safety mesh below. He told the 'Evening Gazette': " It was really just like a scene from a farce. When I saw water underneath me I almost had a fit of hysterical laughter because it didn't seem possible. I was absolutely flabbergasted, my first thought was how can this be?" It was fortunate that poor old Terry gained nothing more than a bump on the head but a serious aspect of the 'farce' was that the barriers had apparently been left raised.

In 1983 Dryboroughs brewed a strong, dark beer specifically for Teesside. They launched the beer on the Bridge using the Transporter in the beer's motif as a symbol of strength and identity.

In October 1986 the Bridge celebrated its Seventy-Fifth Anniversary. To mark the occasion Matthew Walton and Stanley Green completed a ten-foot long scale model of the Bridge. It was a fully-working replica of the Bridge and was itself an achievement of fine precision which reflected the gentlemen's great patience. It was displayed in an exhibition in the Dorman Long Museum in Linthorpe Road.

Terry Scott following his crash - March 1974

13 HOW THE BRIDGE WORKS

The 'transporter' element in the Bridge is, of course, the travelling car. The passenger car is suspended by 30 wire ropes from the upper carriage which is supported on the bottom chord of the trusses. This upper carriage, or travelling frame, is moved by an endless rope system from the electrically powered winch in the winch house on the Middlesbrough side. A description of how the Bridge works has been given by Garth Drabble, formerly the County Surveyor and Engineer of Cleveland: "The winch is powered by two 30 hp motors using a DC electric supply which is first carried up one of the towers and then along the length of the Bridge by means of bare trolley wires, then down to the driver's cabin; in here there are two regulators to start and stop the carriage and regulate the speed. The tram controllers which control the speed of the car are not in the same condition as the motors, and with such large currents being used the copper contact tips which regulate the current have to be constantly maintained. There is a further identical regulator in the winch house and this can be used to bring the carriage back to shore should a fault occur on the overhead power system.

The upper moving carriage has fitted to it 60 wheels and these run across four lengths of rails fixed to the boom.

Slung directly underneath is the lower carriage and this is suspended by 30 steel cables. The conductor directs vehicles onto this lower carriage then signals to the driver who, after assuring himself that the river is clear of shipping, proceeds to cross to the opposite side."

The crossing takes just over two minutes. Cleveland Transit Ltd. is responsible for the operation of the Transporter. The operating team of about seven men work on a shift basis to cover the Bridge's hours of operation:

5 am - 11 pm Monday - Saturday; and 2 pm - 11 pm on Sunday.

There was formerly a tollbooth on each of the approaches to the Bridge, but now fares are collected by conductors on the car deck. In order to allow room for the conductors to move freely the number of vehicles allowed on the deck is now limited to nine.

The Bridge can carry over 600 passengers, or 850 with no vehicles, and in 1988 was estimated to carry 750 foot passengers and 600 vehicles a day.

This cantilever truss type bridge is still structurally in very good condition, thanks largely to the original painting of the Bridge with two coats of red lead priming paint. Today red lead is no longer used due to its high toxic content and drying time of two weeks per coat. What is now used is a system of patch primed High Build Zinc Phosphate with an alkyd gloss finish which provides the essential protection to the steel framework.

The Bridge carries navigation lights, and constant contact is maintained with the Harbour Authority for advance warning of any approaching shipping. Radio contact is maintained by all operating staff to ensure effective communication and safety.

Contact is also maintained with Tees-side Airport and the Newcastle Weather Centre for warnings of fog or high winds. The Tees is prone to occasional dense fog and operating in such conditions is done with great caution - with lights on and regular sounding of the warning bell. The hazard of strong winds is strictly monitored and the Bridge is not operated in winds in excess of 45 mph.

The effective running of the Bridge demonstrates why indeed this design was chosen and is still the most appropriate for this crossing. The height of the Bridge allows for the passage of large shipping without the inconvenience that a swing-bridge would cause during periods of regular river traffic.

A more detailed account of structural dimensions, operation and maintenance is given in the History of the Transporter Bridge written by Garth Drabble and printed privately by the County Council in 1988.

14 OTHER TRANSPORTER BRIDGES

The world's sixteen transporter bridges were all built within a period of 23 years.   The first such bridge, built by Arnodin was the Puente Vizcaya.   Referred to by most as the Bilbao Transporter it is six miles from the city near the mouth of the River Nervion at Portugalete (Bilbao).   Built in 1893 it is believed to be still in action today.

For a complete list of transporter bridges worldwide see Section 20 on page 33.

Suspension Bridge, Bilbao

The first "transbordeur" to be opened in France was at Rouen and went into operation in September 1899.   Unfortunately it only lasted for 40 years.   In June 1940, like the French Fleet it was sacrificed in order to hinder the German military and logistic network, and was blown up by French Army engineers.

The bridge at the naval base at Bizerta in Tunisia lasted only a decade before being dismantled but was reassembled at the Brest naval base in France.   This bridge, too, was a victim of World War Two and was dismantled in 1947 after being irreparably damaged during 1944.

The transporter bridge design was also used in a German naval base at Kiel.   Built in 1909 it linked the base with operations the other side of a deep inlet where the base was extended.   However, this extension to the base was no longer used following World War One; the bridge became an obstacle to large shipping and was closed in 1923.   Because of the prominent roles taken by the Frenchmen Arnodin and Imbault in the history of transporter bridges, the design has been largely the preserve of Europe.   However, two bridges were built outside the Continent: at Duluth, Minnesota, USA, and at Rio de Janeiro in Brazil - the Ponte-Pensil Alexandrino de Alencar.

Cable Suspension Bridge (diagonal anchorages), Rouen

The Widnes to Runcorn bridge, the widest and lowest of all such bridges, was dismanted in 1961. The private bridge of Joseph Crosfield & Sons, also across the Mersey, at Warrington, still exists but is no longer in operation. Meanwhile the Newport Transporter in Gwent only survives because of scrapping costs.

Other transporter bridges have been proposed but never constructed, at Tancarville on the Seine south of Rouen, and at Garonne, just south of Bordeaux. In England similar proposals included a bridge to link Poole with Studland in Dorset, in 1905, and one to link Eastney and Hayling Island in Hampshire.

More interesting was the proposal to build a transporter bridge across the River Liffey in Dublin in 1928. Mr. S.E. Burgess, the County Engineer who supervised the construction of the Middlesbrough Bridge, was consulted by the Dublin Port and Docks Authority regarding the proposal to construct a twin-car design which would have twice the loading capacity of the Middlesbrough Transporter - an estimated 860 vehicles an hour. He estimated that the cost of such a bridge would be approximately £200,000, which was a likely factor in the dropping of the proposal.

The Middlesbrough Transporter is most certain to be the last surviving of the UK bridges, and perhaps will outlive the other transporters around the world.

Cantilever Suspension Bridge (vertical anchorage), Marseille

15 THE BRIDGE'S FUTURE - TOWARDS EXTINCTION OR RENAISSANCE ?

From the 1970s onwards the future of the Bridge has frequently been in question. Earlier indications of its profitability began to give way to doubts about its financial viability. By the end of the 1930s it was estimated that the bulk of the cost had been paid for and in 1961 the Bridge was still able to operate at its original tariff rate of 1d per passenger and 6d per car. With over 1.5 million passengers a year operators stated that the Bridge was running at a clear profit costing 3s 6d each trip with an average revenue of 4s 7d per crossing.

It was not until the Bridge was 57 years old that the first toll increases were announced. In July 1967 the tariff for a passenger doubled from 1d to 2d. The fare for vehicles weighing over a ton rose from 6d to 8d while the charge for those weighing less increased by a penny to 4d. The fare for a cart remained at 6d, each extra horse or mule incurring an extra 2d. An ox or a cow was charged 3d and a sheep or goat would be charged one penny. Revenue from goat traffic was not anticipated to be substantial.

Confidence in the Bridge's future was demonstrated when, in 1969, oxygen pipes were laid over the bridge to link the supply to industrial users on the north side of the river. In June 1970 N. Forbes of the Light Railway Transport League predicted that the Middlesbrough Transporter Bridge would be the last surviving bridge of its kind in the world and it was estimated that it had at least fifteen more years of use in it.

However, during the seventies due largely to the effects of strikes, breakdowns and the cost of repairs, doubts about its future began to loom and talk of scrapping it was heard on several occasions. It was less than encouraging to hear that the Newport Transporter Bridge in Gwent was saved from such a fate only because of the estimated scrapping costs - a rather negative form of reprieve. As pessimism grew supporters of the Bridge sought assurance.

Harry Mead of the 'Northern Echo' has, over the years, written several well-informed and committed articles concerning the Bridge. A loyal devotee of the Transporter, Mead wrote an impassioned article which appeared in the Northern Echo on 23rd September 1976. He wrote: "People living beyond the area might find it hard to believe that no one spoke up immediately for the long-term prosecution of the Transporter. In particular, no Councillor apparently thought it is in the Bridge's favour that it is virtually the only building in Middlesbrough that might tempt an outsider to visit the town."

This was a valid point and while the article was highly critical of Council attitudes of the time it demonstrated Mead's commitment to articulating public affection for the Bridge. Added credence was given to his points about Council indifference when a proposal to illuminate the Bridge at night was rejected in 1978, one Stockton Councillor calling it a 'ridiculous idea'.

The Transporter has undeniably had its problems but these have served only to add to its character, and its imperfections over the years have lent the Bridge a personality rather like that of an ageing aunt. It is not possible to imagine a panoramic view of Teesside without the skeletal structure of the 'Tranny'. It can be seen from so many distances and angles that it is much more than a logo on Council notepaper. It is a focal point of local pride and a symbol of achievement and hope. As I write during a recession it is difficult to see how anybody would wish to destroy such a symbol.

In May 1986 an article in the 'Evening Gazette' titled 'A Bridge Too Far' displayed the re-emergence of the whole question but confirmed that scrapping the Bridge was a 'non-starter'. In 1991 Councillor David Walsh of Cleveland County Council, and Alan Timothy of Teesside Tomorrow, appeared in the 'Evening Gazette' with the appeal 'Don't let this landmark become extinct.' It is a source of great comfort that this is now the overwhelming sentiment within Clevland County Council who run the Bridge through the operators, Cleveland Transit. There is a genuine commitment which augurs well for the Bridge's future and plans are afoot to promote it as a tourist feature.

In September 1987, a report to Middlesbrough Borough Council's Industrial and Commercial Policy sub-Committee highlighted the economic importance of tourism and the role it could play in the strategic thinking behind Middlesbrough's local economy. The report acknowledged Middlesbrough's limited tourist appeal and identified the Transporter Bridge as the town's greatest single attraction. The Middlesbrough Landscape Plan suggested that in order to promote the Bridge as a tourist attraction the surrounding area required substantial development in order to increase the potential for a genuine tourism drive. In May 1990 the existing environs were evaluated and it was decided that a Master Plan was required in order to form the basis of a regeneration strategy. In May, 1991 the Newcastle-based architects, The Napper Collerton Partnership, were commissioned by Middlesbrough Borough Council to produce a proposal for the scheme.

The proposal, published in October 1991, included the construction of a visitor centre on the south side of the Bridge immediately next to the Bridge where the fire boat service is presently based. This would house local tourist information and sales area, an exhibition on the Bridge's history, cafe, bar and toilet facilities and a steward's flat for 24 hour security. Original proposals also included stairs and a lift to a viewing platform and a control room for the Bridge's illumination.

The proposal also concentrated on the immediate vicinity of the Bridge: the area from Albert Road through Queens Square and Cleveland Street up to Durham Street, East Street and Ferry Street. This included an increased car parking area, upgraded street lighting and repaving.

The plans demonstrated the potential for aethsetic development of the locality and included the creation of boulevards of which the Parisian town planner Haussmann himself would be proud. They certainly provided the focus for effective long-term planning and an environmental commitment to tourism, local industry and investment and the local community of St. Hilda's. The strategy relied upon the co-operation and investment of the public and private sectors. Led by Middlesbrough Borough Council negotiations have included Cleveland County Council, Tees and Hartlepool Port Authority, The Teesside Development Corporation and St. Hilda's Community Council. Consultation with local industry has also been a key element in potential investment and development of land presently owned by private companies such as Trafalgar House and SLD Engineering Ltd.

The proposals represent the Council's ambitious attitude towards tourism. The Tourist Section of the Council estimated in 1991 that 30,000 people visit the Bridge each year. As part of the Council's Landscape Plan the Bridge would be the pivotal feature in the wider development of tourism. The Transporter provides a physical link with the International Nature Reserve at Seal Sands, the seafront at Seaton Carew and Hartlepool Marina. South of the river the Master Plan outlines initiatives to extend the public's access to the river bank beyond Teesaurus Park and the Bridge itself is the culmination of the Town Trail which includes the town's most historical buildings.

The central theme of the Bridge has provided the Council with an opportunity to display its optimism and positive intent towards the town's economic and environmental future.

16 BREAKDOWN TO REPAIR

At approximately 8 am on the 20th March 1990 a running wheel fell in front of a motorist's windscreen onto the car deck. The driver was unscathed and minimal damage was caused to his vehicle, but the event certainly woke him up. The wheel's descent was broken, fortunately, by supporting struts and safety mesh on its way down but it was a matter of great chance that nobody was seriously injured or worse.

The incident brought the local reporters scurrying to the Bridge. It was certainly a shocking episode for the passengers on the car deck, particularly the two in between whose vehicles the wheel landed. So what happened that morning? The maintenance team held an inquiry into the incident against a background of rising public interest.

As the car was travelling across the river the pin in one of the middle wheels in a set of fifteen sheared off. As this sheared the wheel jammed and the two following wheels therefore jammed; their pedestals collapsed and the three wheels came off the rails. Two fell onto the staging boards whilst the third fell onto the moving car deck.

Several girders were twisted as a result of the accident and one of the initial jobs was to repair the girders by putting in stiffeners. The pin that had sheared off was then given to Teesside Polytechnic (now the University of Teesside) for their Department of Mechanical Engineering to carry out tests. The incident at least provided valuable experience for a group of engineering students through a stimulating and high profile local project. The Polytechnic, having run thorough tests on the metallurgical fabric of the metal, produced a report which concluded that the primary cause of the accident had been metal fatigue on the axles. The next stage was a series of magnetic particle tests on all the remaining wheels and pedestals, which took considerable time.

By August 1990 the Bridge was working but for maintenance purposes only. During the autumn the repairs to wheels and pedestals began to move swiftly. The pins on every wheel and all the pedestals were totally renewed. A local firm, Rockliffe Engineering, set about manufacturing new pedestals and pins. New holes had to be drilled on the three girders which were repaired having been severely twisted. The wheels were fitted with what are effectively steel tyres, heated and shrunk onto the wheel and re-profiled.

A part of the wheel that fell onto the car deck - March 1990

In January 1991 the maintenance team set about replacing all the rolling gear and carried out test runs - proving runs - of the car. As all the initial repair work had been completed at once the running was very tight and the power produced very high amp readings. An attempt was made to relieve the loading on the rails by increasing the running surface of the wheels and in effect simulating the natural wearing-in process. This was facilitated by increasing the width between the flanges on the wheels from 2 1/8" to 2 1/4". This did slightly improve the running as it reduced the tightness of the moving parts. The team began switching pedestals and re-arranging the wheel combinations as they explored further avenues in trying to bring the power readings down.

In March 1991 it was decided that outside consultants should be brought in. An investigation was completed by the Offshore Certification Bureau consortium and a report initiated the following May. The report and recommendations were presented in September 1991. Following appeals for funds towards the recommended remedial work, the firm Rendel Palmer & Tritton, the member of the OCB consortium most closely involved in the initial investigation, was appointed to direct the work of the maintenance crew in implementing the recommendations.

During this time the operating staff were employed replacing all the nuts and bolts along the rails where they were needed. Most of them had last been replaced during the Bridge's closure in 1977. These replacements and proving runs will form the backbone of a constant maintenance programme that will continue as long as the Transporter remains in operation.

The closure of the Bridge provided Cleveland County Council with the opportunity to make large scale repairs to the Bridge, not merely to the specific parts involved in the incident. The complete overhaul was necessary for a bridge of this age and the extended closure for repair was an encouraging indication for the confidence in the Bridge's future safety and efficiency. It was also a demonstration of the Council's commitment to the Bridge's long-term future.

Alan White, of the Bridge Maintenance Department of Cleveland County Council, is optimistic about the Bridge's future. He has also identified the Bridge's tourist potential and recalls showing a fascinated American engineer around the car. He is confident that the Bridge will continue to operate for many more years and again thrive as a vital link across the Tees.

Ruth English, Middlesbrough Tourism Officer, shows the Transporter to tourism chiefs from throughout the North - 14th July 1992

17   DEVELOPMENTS SINCE 1992

Replacement of the Rails

The rails along the top of the Bridge had been replaced in 1977 with a 2 mm gap at the joints and a conventional four-hole fishplate. However, a survey in September 1997 revealed an extensive pattern of failures with 42 of the 166 rail ends inspected affected by star cracking of the fishplate holes while 86 were cracked at the upper fillet at the rail head and web interface. In May 1998 the consultant recommended:

- *A heavier rail section to eliminate fishplate joints by welding*
- *If fishplate joints were to be retained the cold expansion pre-treatment of fishplate holes would be necessary and the provision of six-hole fishplates*
- *Renewal of the rail pad was required*

P.C. Richardson of Middlesbrough was appointed in December 1998. Work was completed in June 1999 at a cost of £325,000. The materials used were:

- *BS40 flat-bottomed rail*
- *Six-hole fishplate*
- *Rail pad*

Energy Saving Floodlights

The Bridge's winter floodlights were switched on by Councillor Ron Lowes, Executive member of Transport and Housing, at 4.15 on Thursday 3 November 2006. He said, *'This year we've secured funding of £42,318 from the 'Invest to Save Fund for Energy Conservation' which means we've been able to replace the original floodlights with 46 new fittings that give an energy saving of 24 per cent. This has resulted in a yearly cost saving of nearly £1,200 and a carbon saving of 13,183.8 kg of carbon dioxide per annum which is good news for us and for the environment.'*

Fractured rail

Winding gear

Painting of the Bridge

The Bridge was painted in 2003. The successful contractor was T.I. Protective Coatings from Bolton. The scheme cost £535,000 and painting was done by rope access.

18   NEW TECHNOLOGY FOR A NEW CENTURY

In 1911 the Bridge was at the leading edge of engineering technology at a time when the Tees Valley was dominated by iron and steel and engineering companies. The region is still at the leading edge of technology but today the emphasis is on innovation in alternative fuels. And in 2004/05 the Transporter Bridge was home to one of the first real-world applications of hydrogen fuel-cell technology. The trial was carried out by Renew Tees Valley and the Centre for Process Innovation based at the Wilton Centre. The

idea was to use a hydrogen fuel-cell to power a sign that would advise motorists on the approach to the Bridge whether it was open or closed. The challenges faced in a real-world environment compared with a laboratory were: designing for safety in an environment where members of the general public (Bridge passengers) and sources of ignition (e.g. cigarettes) were present; coping with high electricity demand from the sign in bright weather; coping with water vapour condensation in cold weather etc. People came from Korea, Japan, Germany, Australia and the USA to see how we had done it. The learning from the Bridge has been used to develop a system elsewhere in Middlesbrough which is 50 times bigger and to start work on a design which is 200 times bigger again.

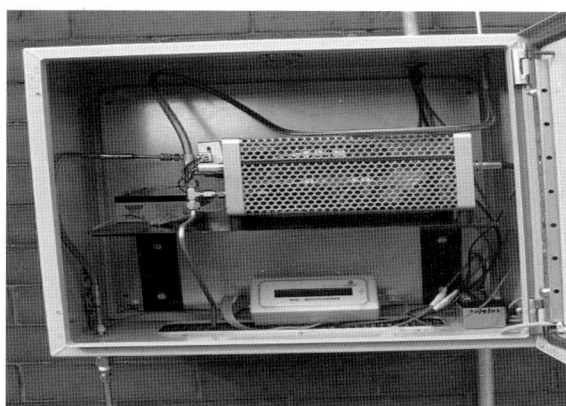

The hydrogen-powered fuel-cell

Close up of the fuel-cell powered sign

19   THE VISITOR CENTRE

This was opened by the late Fred Dibnah in June 2000. Work started in February to replace an existing building with the demolition of the existing workshop and the reclamation of existing wrought-iron fencing. The blacksmith at Preston Park Museum in Stockton recreated additional railings to enable the formation of the external viewing area. Since the existing workshop was a listed building the size, materials and appearance of the new Visitor Centre were limited. The work was carried out by a local contractor SGW Construction. The Visitor Centre and the Bridge are now used on a regular basis by organisations for charity fund raising events which include walks up the Bridge, abseils, zip slides and bungee jumps. Each year the Bridge is open in September for a nationwide Heritage Day. Members of the public climb up the Bridge to the top walkway and view the winding house. In 2006 a total of 671 people visited the Bridge.

The Visitor Centre is used throughout the year for meetings, presentations and seminars. Schools use the Centre for lessons relating to the river and Bridge structure. The Visitor Centre is open Monday - Saturday from 9am to 5pm and on Sundays from 2pm to 5pm.

Fred Dibnah (centre) with Brian Glover (left)
Head of Transport & Design Services and Rodger
Wakerley, Principal Engineer, Middlesbrough Council

20 TRANSPORTER BRIDGES WORLDWIDE

• Bilbao (1893)** Spain	• Brest (1909) France
• Bizerta (1896) Tunisia	• Kiel (1909) Germany
• Rouen (1899) France	• Osten (1909)** Germany
• Rochefort-Martrou (1900)** France	• Middlesbrough (1911)** Britain
• Nantes (1903) France	• Rendsburg (1913)** Germany
• Marseille (1905) France	• Rio de Janeiro (1915) Brazil
• Duluth (1905)X USA	• Buenos Aires (3- 1913/1914/1915)** Argentina
• Widnes (1905) Britain	• Warrington (1916)** Britain
• Newport (1906)** Britain	• Bordeaux (never completed) France

X Transformed into a lift bridge
*** Still in operation, some for occasional tourist purposes only*

In September 2003 a meeting was held in Bilbao, Spain, of representatives of the transporter bridges still in existence (see table).

The meeting was held to discuss the possibility of forming a World Association of Transporter Bridges with a view to applying for European funding. Each of the bridges' representatives agreed that it was an excellent idea, however, the legal aspects of the Association had to be reviewed by each group.

In May 2005 the representatives met again in Rochefort in France and it was agreed that the way forward was for each country to form their own association of transporter bridges with a view to apply to the European Community as a group.

In August 2006 the local authorities of Newport, Warrington and Middlesbrough signed an agreement and formed the British Association of Transporter Bridges.

The King of Spain talking to
Councillor McPartland and
Rodger Wakerley